THE CHANNEL TUNNEL

Alan Trussell-Cullen

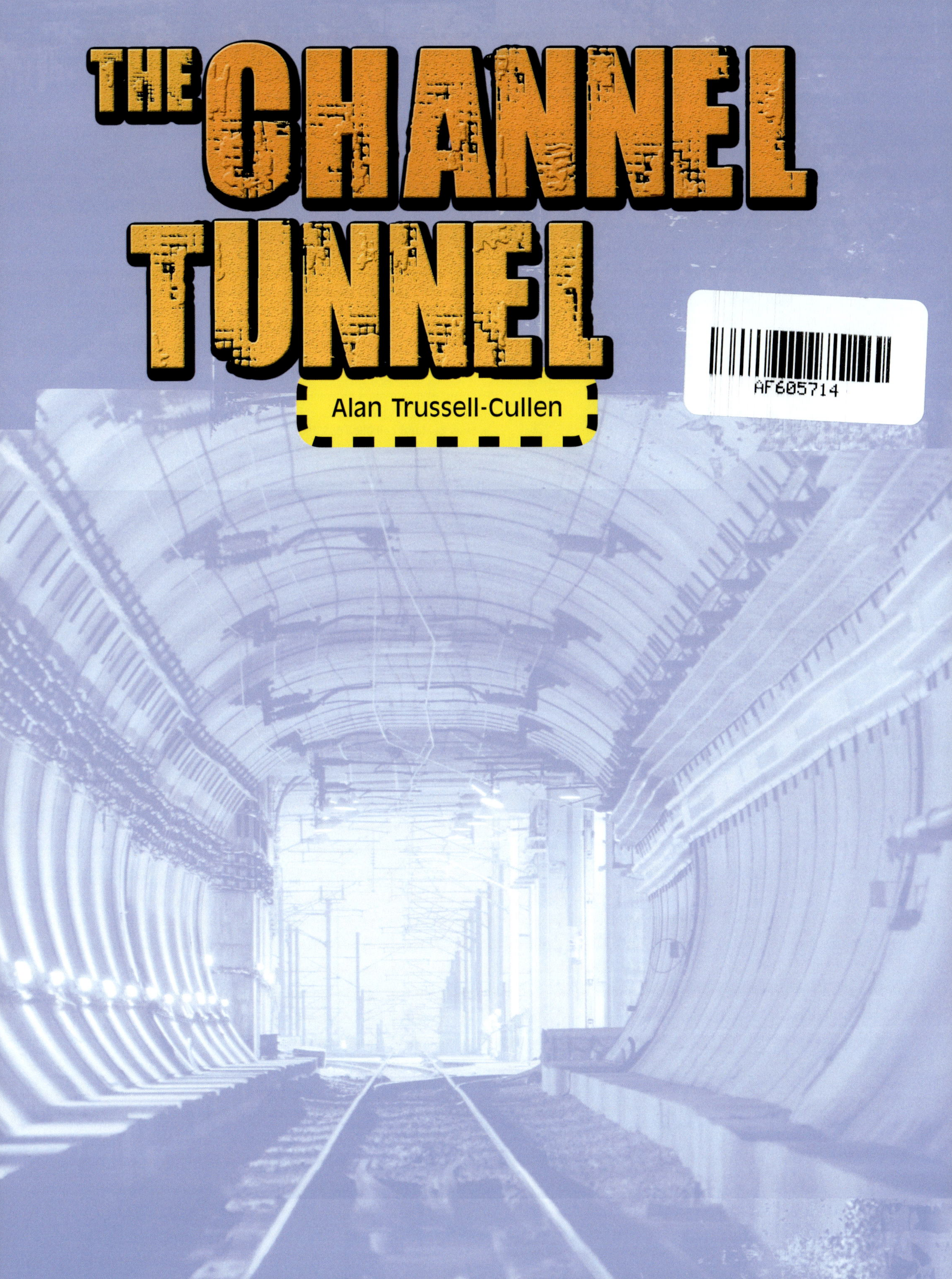

The Channel Tunnel

Text: Alan Trussell-Cullen
Editor: Ben Haskin
Design: Jennifer Warwick
Series design: James Lowe
Photo researcher: Lisa Piemonte
Production controller: Adam Bextream
Reprint: Siew Han Ong

Acknowledgements
The author and publisher would like to acknowledge permission to reproduce material from the following sources:
Alamy/David R. Frazier Photolibrary, Inc.: p. 17 (main); Alamy/G P Bowater: p. 21 (main); Alamy/qaphotos.com: pp. 1, 5, 7, 17 (inset), cover, back cover; Corbis/Nogues Alain/CORBIS SYGMA: p. 22; Corbis/Stapleton Collection: p. 8; Eurotunnel: pp. 20, 21 (inset), 23; Getty Images: pp. 3, 9 (main), 9 (inset), 18; Julian Bruère © Cengage Learning Australia: p. 10; Jupiterimages Corporation: p. 13; PA/EMPICS: p. 19; PA/PA Photos: p. 15 (inset), 15 (main); Photolibrary/Mary Evans Picture Library: pp. 12, 14; Photolibrary/Sheila Terry: p. 11; Shutterstock/Christoffer Vika: p. 16.

Every effort has been made to trace and acknowledge copyright. However, if any infringement has occurred, the publishers tender their apologies and invite the copyright holders to contact them.

Fast Forward Independent Texts
Level 25

For product information and technology assistance,
in Australia call 1300 790 853;
in New Zealand call 0508 635 766

For permission to use material from this text or product,
please email **aust.permissions@cengage.com**

ISBN 978 0 17 017980 5
ISBN 978 0 17 017899 0 (set)

Cengage Learning Australia
Level 7, 80 Dorcas Street
South Melbourne, Victoria Australia 3205

Cengage Learning New Zealand
Unit 4B Rosedale Office Park
331 Rosedale Road, Albany, North Shore NZ 0632

For learning solutions, visit **cengage.com.au**

Printed in Australia by Ligare Pty Ltd
3 4 5 6 7 23 22 21 20

THE CHANNEL TUNNEL

Alan Trussell-Cullen

Contents

CHAPTER 1

What Is the Channel Tunnel?

The Channel Tunnel is a railway tunnel that joins England and France. It goes under the English Channel, the sea between the two countries. It is also known as the "Eurotunnel" or the "Chunnel". Some people think that the Channel Tunnel is one of the wonders of the modern world.

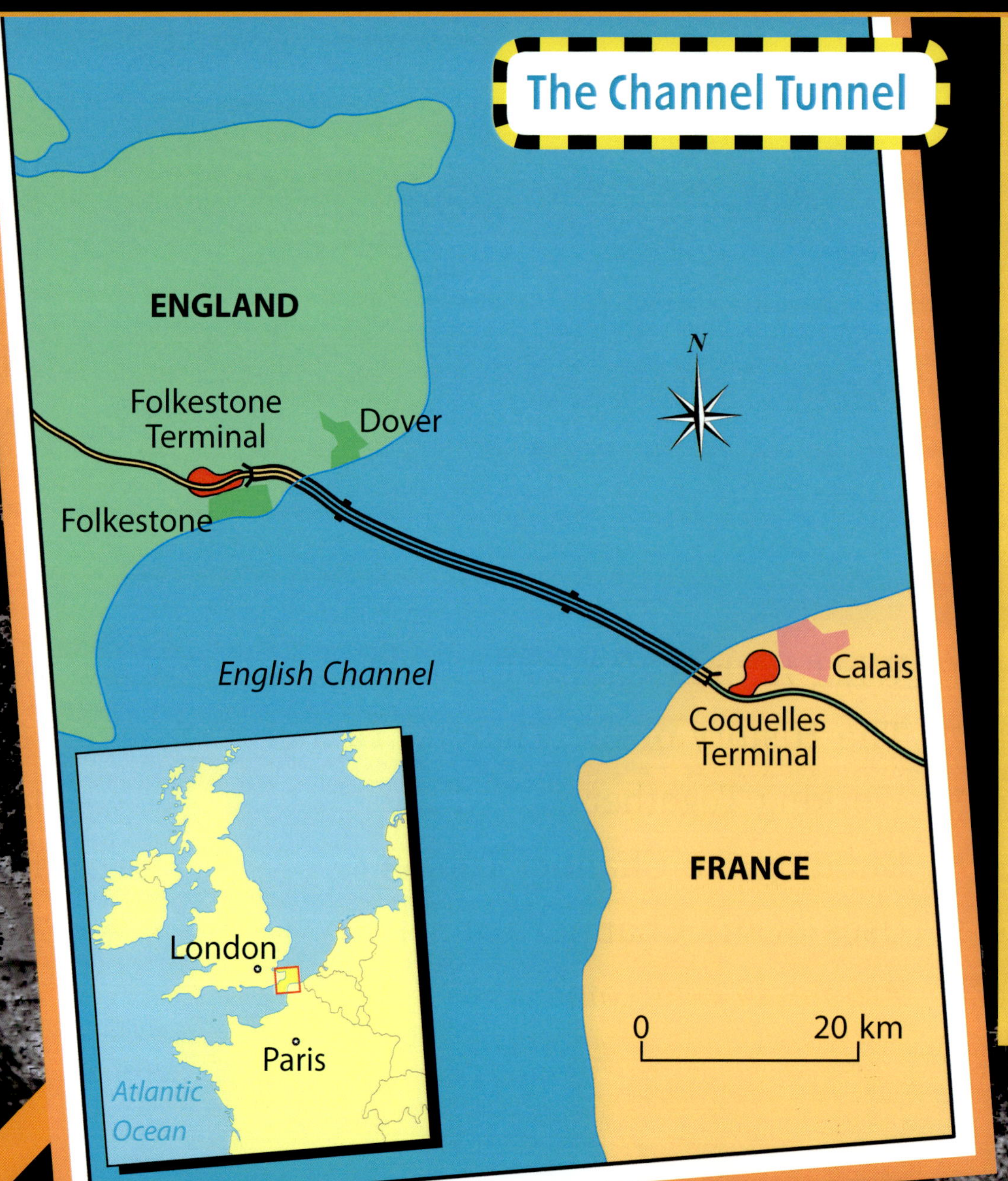

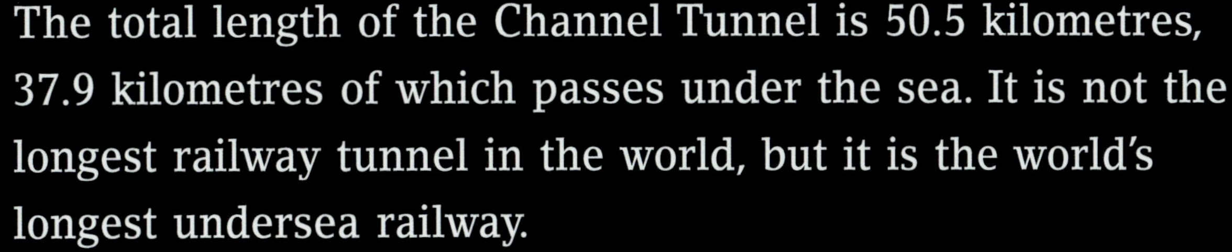

The total length of the Channel Tunnel is 50.5 kilometres, 37.9 kilometres of which passes under the sea. It is not the longest railway tunnel in the world, but it is the world's longest undersea railway.

Channel Tunnel Fast Facts

Year completed:	1994
Time to complete:	6 years
Length:	50.5 kilometres
Depth below sea bed:	40 metres to 75 metres
Number of workers:	about 13 000 people

The Channel Tunnel is actually made up of three main tunnels. There are two railway tunnels, one for trains going from England to France and the other for trains going from France to England. A smaller service tunnel runs between the two railway tunnels.

The Channel Tunnels

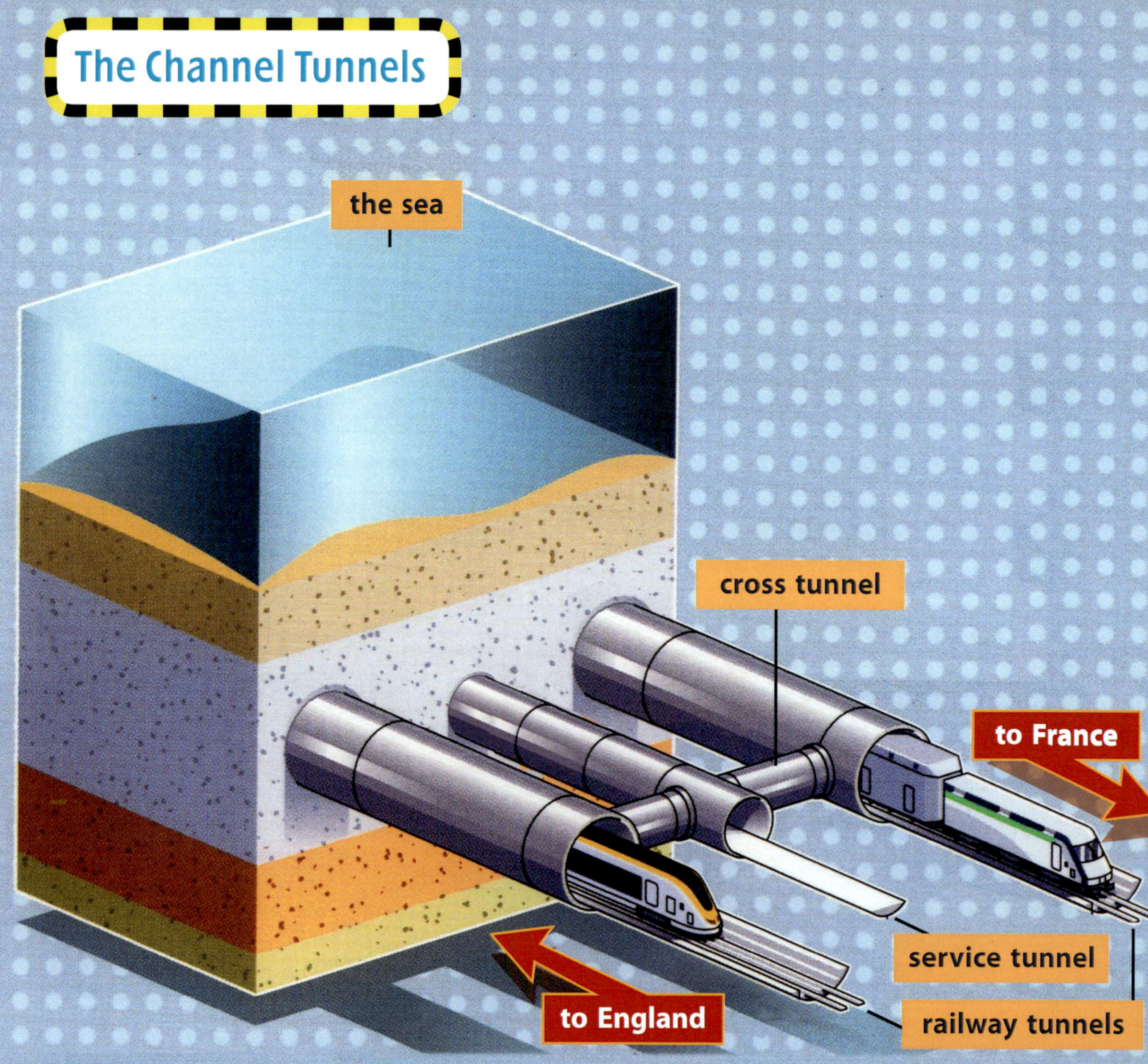

a cross tunnel seen from one of the railway tunnels

Small cross tunnels link the two railway tunnels with the service tunnel. Workers use these cross tunnels to move between the service tunnel and the railway tunnels when something needs to be repaired.

The cross tunnels can also be used in an emergency to allow people to escape safely from the railway tunnel into the service tunnel.

CHAPTER 2

The History of the Channel Tunnel

For centuries, the only way people could travel between England and France was to cross the English Channel by boat. The English Channel is a very dangerous body of water in stormy weather. Over the centuries, hundreds of ships and boats have been shipwrecked and sunk and many people have lost their lives.

a shipwreck on the English Channel

Many people have dreamed of other ways to cross the Channel. People had ideas for bridges, and the first successful crossing by hot-air balloon occurred in 1785.

an idea from 1906 for a bridge across the Channel featuring a monorail

An artist's idea of what the balloon crossing in 1785 might have looked like.

In 1802, French mining engineer Albert Mathieu presented the Emperor of France, Napoleon Bonaparte, with an idea to build a tunnel under the Channel. The tunnel would allow people to travel between France and England more easily. Mathieu imagined that people would travel through the tunnel in horse-drawn coaches, by the light of oil lamps.

But Mathieu faced a number of problems. He did not know what sort of rock the sea bed was made of, or how difficult it would be to tunnel into it. Many people in England did not support the plan as they were worried that France might use the tunnel to invade England. In any case, at that time the technology did not exist to build the tunnel.

The plan was abandoned when war broke out between France and **Britain** in 1803.

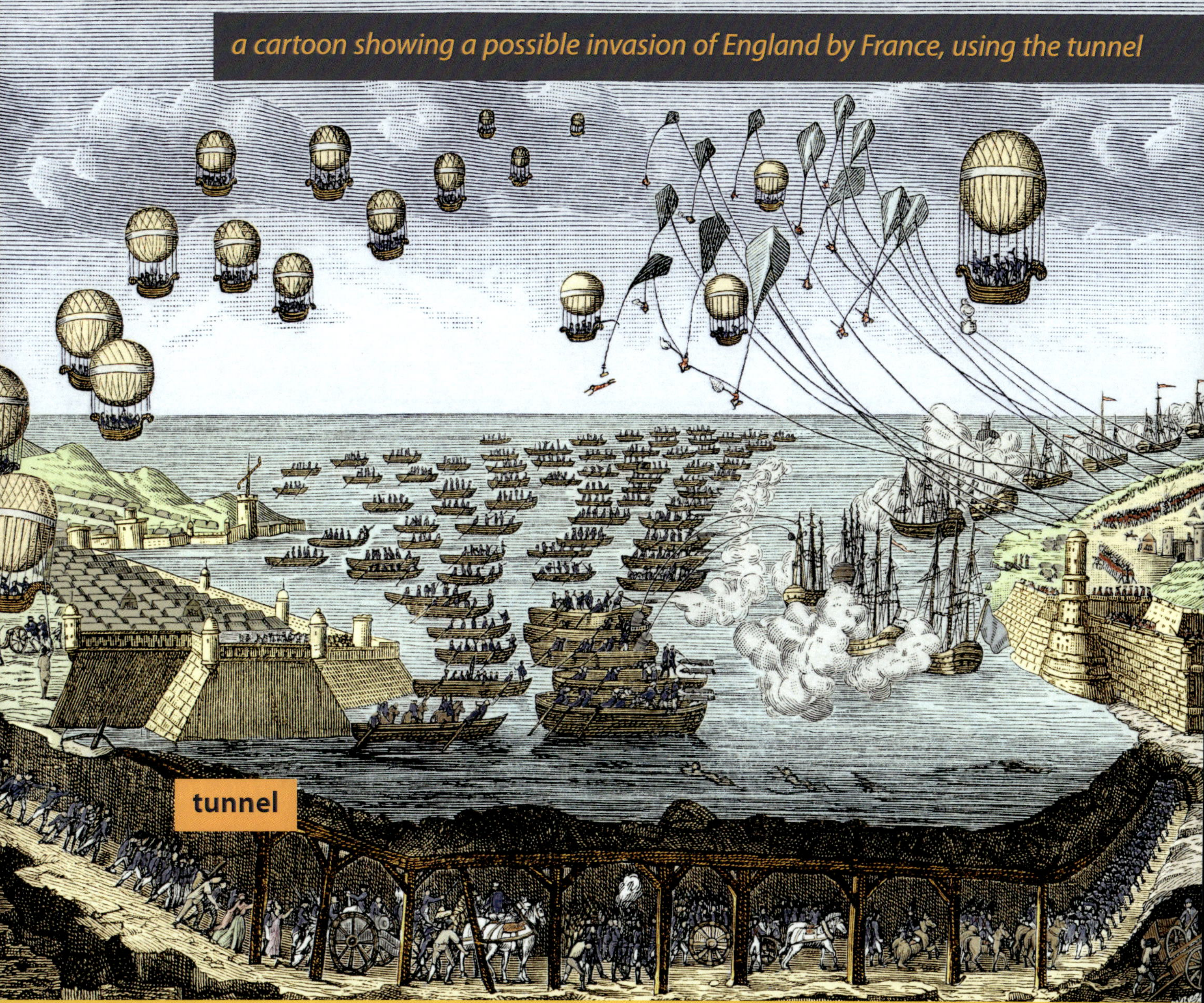

a cartoon showing a possible invasion of England by France, using the tunnel

In the nineteenth century, steam engines were being used to power trains, and railways were being built all over the world. Steam trains made it easier to travel between England and France. But people still had to cross the Channel in small ferries, and this was the most dangerous part of the journey.

crossing the Channel by ferry

As train travel became more popular, engineers started to think about the idea of a tunnel under the Channel once more. They imagined a railway running all the way from London in England to Paris in France.

But some of the problems that Mathieu faced had not been solved. The engineers still did not know if the rock in the sea bed was suitable to tunnel into.

Also, the steam trains created a lot of smoke which would be trapped in the tunnel. The smoke would be a health hazard, and passengers might choke to death.

CHAPTER 3

Working Together

In the 1870s, the governments of Britain and France decided to work on a plan for a tunnel together. They both set up tunnel companies.

Engineers and scientists explored the sea bed to find out what sort of rock it was. They decided to dig the tunnel in a thick **chalk** bed. The English and French tunnel companies began **boring** trial tunnels in 1881, and a year later, almost two kilometres had been bored from each end.

But the English still feared a French invasion, and in 1883, further building was banned.

workers in an early trial tunnel

By the end of the 1800s, the problem of the smoke and fumes given off by steam trains had been solved – electric trains had been invented.

Over the next hundred years, there was more talk of the tunnel, but it wasn't until 1976 that the two governments finally agreed to another attempt.

an electric train from the early 1900s

1976: British Prime Minister Margaret Thatcher and French President François Mitterrand agree that their countries will work together on the Channel Tunnel.

CHAPTER 4

Building the Channel Tunnel

During the 1980s, scientists and engineers **confirmed** that most of the rock in the sea bed was chalk. This was good news because chalk is easy to bore through but also stands up well on its own.

The white cliffs of Dover are made of the same chalk that is under the English Channel.

a tunnel-boring machine

In 1987, work finally began. By then, many similar tunnels had been built around the world, so the technology to build the Channel Tunnel was available. Huge tunnel-boring machines were used to bore the tunnel from each end, in England and France.

On 30 October 1990, the machines boring the service tunnel met in the middle of the tunnel. The machines boring the two railway tunnels broke through in May and June 1991.

About 13 000 people worked on the tunnel, and it was completed in six years. The Channel Tunnel was opened by the British **monarch** Queen Elizabeth II and the French President François Mitterrand at a ceremony in Calais in France on 6 May 1994.

CHAPTER 5

Uses of the Channel Tunnel

There are four different types of trains that travel through the Channel Tunnel.

- Eurostar trains are high-speed, long-distance passenger trains. They travel between London and Paris in two and a half hours.

- Eurotunnel shuttles carry cars, buses, vans and caravans in special railway **carriages**. People drive on at terminals near Folkestone, England, or Calais, France. They stay in their vehicles for the trip, and then drive off at the other end.

- Eurotunnel **freight** shuttles carry trucks that drive onto open railway carriages at the terminals. The drivers then ride in separate passenger cars.

- Freight trains carry freight and containers.

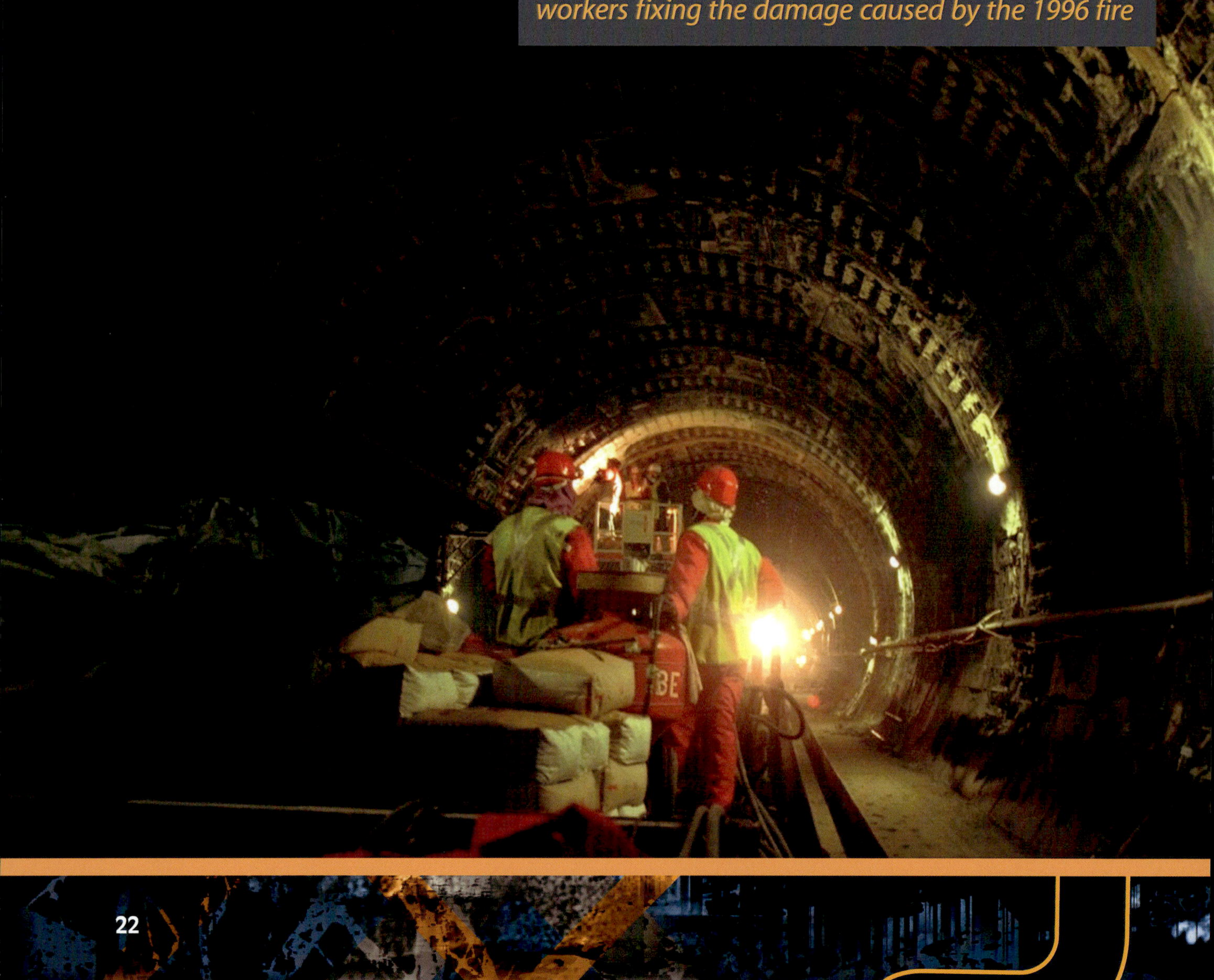

workers fixing the damage caused by the 1996 fire

Despite these problems, the tunnel has many modern safety systems and is a very safe way to travel.

The Channel Tunnel now provides a quick way to travel between England and France. Most people agree that the Channel Tunnel is a great **engineering feat**.

This family has driven their car onto a Channel Tunnel train.

Glossary

boring making holes by drilling

Britain the nation that is now made up of England, Scotland, Wales and Northern Ireland

carriages railway cars that are joined to form trains

chalk a soft white rock

confirmed found something to be true

engineering the use of science to design and build things

feat a great achievement

freight goods carried on a train, truck, ship or plane

monarch a supreme ruler such as a king or queen

Index